I0475745

U.S. Department of Justice
Office of Justice Programs
National Institute of Justice

RESEARCH REPORT

Reducing Gun Violence

EVALUATION OF THE
INDIANAPOLIS
POLICE DEPARTMENT'S
DIRECTED PATROL PROJECT

NIJ

U.S. Department of Justice
Office of Justice Programs
810 Seventh Street N.W.
Washington, DC 20531

John Ashcroft
Attorney General

Deborah J. Daniels
Assistant Attorney General

Sarah V. Hart
Director, National Institute of Justice

Office of Justice Programs
World Wide Web Site
http://www.ojp.usdoj.gov

National Institute of Justice
World Wide Web Site
http://www.ojp.usdoj.gov/nij

Reducing Gun Violence

Evaluation of the Indianapolis Police
Department's Directed Patrol Project

Edmund F. McGarrell
Steven Chermak
Alexander Weiss

Crime Control Policy Center
Hudson Institute

November 2002
NCJ 188740

National Institute of Justice

Sarah V. Hart
Director

Edmund F. McGarrell is director of the School of Criminal Justice at Michigan State University and adjunct fellow at the Hudson Institute. Steven Chermak is associate professor in the Department of Criminal Justice at Indiana University. Alexander Weiss is executive director of the Center for Public Safety and professor of management and strategy at the J.L. Kellogg Graduate School of Management at Northwestern University.

This research was sponsored by grant award number 95-IJ-CX-0019 from the National Institute of Justice, U.S. Department of Justice. Points of view or opinions expressed in this report are those of the authors and do not represent the official position of the National Institute of Justice or the U.S. Department of Justice. This document is not intended to create, does not create, and may not be relied on to create any rights, substantive or procedural, enforceable at law by any party in any matter civil or criminal.

Foreword

This Research Report is part of the National Institute of Justice's (NIJ's) Reducing Gun Violence publication series. Each report in the series describes the implementation and effects of an individual, NIJ-funded, local-level program designed to reduce firearm-related violence in a particular U.S. city. Some studies received cofunding from the U.S. Department of Justice's Office of Community Oriented Policing Services; one also received funding from the Centers for Disease Control and Prevention.

Each report in the series describes in detail the problem targeted; the program designed to address it; the problems confronted in designing, implementing, and evaluating the effort; and the strategies adopted in responding to any obstacles encountered. Both successes and failures are discussed, and recommendations are made for future programs.

While the series includes impact evaluation components, it primarily highlights implementation problems and issues that arose in designing, conducting, and assessing the respective programs.

The Research Reports should be of particular value to anyone interested in adopting a strategic, data-driven, problem-solving approach to reducing gun violence and other crime and disorder problems in communities.

The series reports on firearm violence reduction programs in Boston, Indianapolis, St. Louis, Los Angeles, Atlanta, and Detroit.

Contents

Introduction

During the mid-1990s, Indianapolis found itself in an unusual situation. The local economy was strong and the city's downtown was experiencing a vibrant renewal. But the city also was experiencing record-setting levels of homicide at a time when homicide was declining in many comparable cities.

Local officials took several steps to address homicide. For example, they used data to identify where and when homicides were occurring. To produce the data, the Indianapolis Police Department (IPD) created the Indianapolis Management Accountability Program, or IMAP, an adaptation of the New York City Police Department's computer comparison statistics (CompStat) program.

IPD then applied directed patrol tactics in two areas of the city that had high concentrations of violent crime. Directed patrol involves assigning officers to a particular area to proactively investigate suspicious activities and enforce existing gun, drug, traffic, and related laws. Officers assigned to directed patrol areas are freed from having to respond to calls for service.[1]

Directed patrol is thought to be most promising as a crime control tool when it is targeted toward high-crime locations and their hot spots.[2] Indianapolis selected the approach because research indicated it had been successful in Kansas City. (See "Findings From Kansas City.")

The most common approach in a directed patrol effort is traffic stops. The strategy generally includes increasing the number of police officers in a given location and the number of contacts with citizens. In theory, intense traffic enforcement should have a general deterrent effect because it increases the threat of detection and punishment of criminal activity.[3]

To the extent that directed patrol focuses on suspicious individuals in high-risk locations, it moves from a general deterrence strategy to a targeted or focused deterrence strategy. The Indianapolis study provided the opportunity to compare a general deterrence with a targeted deterrence strategy.

> ### Findings From Kansas City
>
> One of the most promising studies on the use of directed patrol to reduce violent crime is the Kansas City gun experiment conducted by Lawrence W. Sherman and his colleagues in the early 1990s.* Kansas City police officers, trained to search for illegal guns, increased traffic enforcement in a police beat with high levels of violent crime. Their efforts led to increased seizures of illegal firearms, which in turn were associated with a significant decrease in gun-related crime in the targeted area. The researchers found that the target beat experienced a 65-percent increase in firearm seizures and an approximately 50-percent decrease in the incidence of gun-related crime. A control beat experienced a slight decline in gun seizures and a small increase in gun-related crime.
>
> _____
>
> * Sherman, L.W., J.W. Shaw, and D.P. Rogan, *The Kansas City Gun Experiment,* Research in Brief, Washington, DC: U.S. Department of Justice, National Institute of Justice, 1995, NCJ 150855; and Sherman, L.W., and D.P. Rogan, "The Effects of Gun Seizures on Gun Violence: 'Hot Spots' Patrol in Kansas City," *Justice Quarterly* 12 (1995): 673–693.

How Did Indianapolis Reduce Gun Crime?[4]

IPD applied directed patrol tactics in two police districts in two different ways. Put in the simplest terms, the East District followed a general deterrence strategy whereby it assigned many police officers who stopped many people, issued many citations, and made 1 felony arrest for every 100 traffic stops. The North District, employing a targeted deterrence strategy, assigned fewer officers who stopped fewer people and issued fewer citations but made almost 3 times as many arrests for every 100 stops. Officers in the North District were more likely to stop and arrest felons because they focused on specific suspicious behavior and individuals. Homicide went down in both districts, but the North District also reduced gun crime overall—and they did so using fewer resources.

Directed patrol in the North target area reduced gun crime, homicide, aggravated assault with a gun, and armed robbery. In contrast, in the East target area it had no effect on gun-related crime, except for a possible effect on homicide. Why? The North District's targeted deterrence approach most likely sent a message of increased surveillance to those individuals most likely to commit violent gun-related crimes.[5]

The results of the Indianapolis directed patrol program are consistent with a growing body of research that shows that when police identify a specific problem and focus their attention on it, they can reduce crime and violence. As in the Kansas City gun intervention project, directed police patrol led to sizable reductions in gun crime. Additionally, it did not shift crime to surrounding areas or harm police-community relations.

The finding that the community generally accepted the program supports the idea that crime control benefits need not generate police-citizen conflict. However, the lack of impact in Indianapolis's East target area, which used a more general rather than a targeted deterrence model, and the potential strain that these types of police initiatives could have on police-community relations suggest the need for continued research on both the benefits and the potential costs of such strategies.

About This Report

The evaluation of the Indianapolis directed patrol program examined several questions:

- Can directed police patrol reduce violent gun crime?
- Do different directed patrol strategies have different effects?
- Will the community support this type of aggressive traffic enforcement?
- Which aspects of the Indianapolis experiment work and what still remains unknown?

The following sections address these questions.

Putting the Strategies in Place

The Indianapolis directed patrol experiment was a 90-day project initiated on July 15, 1997, in two districts:

- The North District's beats A51 and A52.
- The East District's beats B61 and B62.

Data from IPD's Indianapolis Management Accountability Program revealed that these four beats were consistently among those that exhibited the highest levels of violent crime, drug distribution, and property crime in the city. (See "Selecting the Target and Comparison Areas.")

The choice of strategies employed to implement directed patrol was left to the command staff of each district. Consequently, each district employed a slightly different strategy.

The East District used a broader general deterrence strategy that maximized the number of police vehicle stops, thereby creating a sense of significantly increased police presence. The strategy was based on the theory that offenders would be deterred by the increased patrols. Additionally, the police anticipated that the large number of vehicles would yield seizures of illegal weapons and drugs.

Selecting the Target and Comparison Areas

To determine the extent to which directed patrol in the North and East target areas had an impact, researchers needed to select a comparison area that did not experience directed patrol.

Selecting comparison areas for evaluations can be problematic. No two areas are alike, and they are influenced by myriad demographic, economic, neighborhood, and police processes. In an ideal situation, the police beats with crime patterns most like those in the target areas would be selected. This process was impossible in Indianapolis, however, because the beats most like the target areas were contiguous to them. Researchers did not want to use contiguous beats as comparisons because they intended to examine crime effects in those surrounding areas.

Demographics

Consequently, two beats in the East District were selected as the most similar available choices (beats B41 and B42). The comparison area, however, is more

Continued on page 6

populous than both of the target areas and covers a significantly larger land area. The comparison area houses primarily black residents (86 percent) and thus is more comparable to the North target area, where most residents live in predominately black, low-income neighborhoods. The low-income neighborhoods in the East target area consist principally of white residents; 14 percent are black and a small but growing number are Latino.

Exhibit 1 presents basic data for the target and the comparison areas, and exhibit 2 presents Uniform Crime Reporting (UCR) Program Index offenses (1996) for the three areas as well as citywide.*

Crime Rate Comparisons

The evaluation compared crime trends in the target areas with crime trends in the comparison area and citywide (minus crime in the target areas). The evaluators assumed that the citywide crime trend would provide the best estimate of what was likely to occur in the target areas absent the directed patrol project.

The homicide rate in the North target area was three times that of the city. Its robbery and aggravated assault rates were almost twice those of the city. Its property crime rate, however, was slightly lower than the city's rate.

The East target area's homicide rate was between the rates of the North target area and the city. It had a particularly high rate of robbery, and its rate of aggravated assault was nearly twice that of the city. Its rate of property crime was higher than the rates in the city and the North target area.

The North and East target areas are quite dense, which reduces their population-based rates of crime. Both areas, however, have extremely high rates of violent crime for their size. Although the comparison area had a higher violent crime rate than the city, its rate was considerably lower than those of the target areas.

---

* Indianapolis is part of a consolidated city-county governmental structure. The police department's jurisdiction consists of the center city. The population in 1990 was 377,723. The crime data and the population base refer to the police department's jurisdiction. The figures differ from those reported in the Uniform Crime Reporting Program, which include the consolidated city-county jurisdiction (population approximately 760,000).

Exhibit 1 Characteristics of Target and Comparison Areas

Characteristic	North Target Area	East Target Area	Comparison Area
Population	16,612	14,645	19,305
Square miles	2.79	1.69	4.74
Person-weeks	215,956	190,385	250,965
Square mile-weeks	36.27	21.97	61.62
Gun crimes (July 15 to October 15, 1996)	75	42	49

Note: Person-weeks are calculated by population × weeks. Square mile-weeks are calculated by square miles × weeks.

Exhibit 2 Level of Crime for Uniform Crime Reporting Program Index Offenses, 1996

Activity	Number of Offenses			
	Citywide	North Target	East Target	Comparison Area
Murder	114	15	7	9
Robbery	2,600	194	229	122
Aggravated assault	4,280	330	301	281
Rape	424	23	25	23
Total violent	7,418	562	562	435
Burglary	7,797	303	564	337
Larceny	16,842	633	796	469
Motor vehicle theft	5,860	295	269	350
Total property	30,499	1,231	1,629	1,156
Total Index	37,917	1,793	2,191	1,591
	Rate per 1,000 Residents			
	Citywide	North Target	East Target	Comparison Area
Murder	0.3	0.9	0.5	0.5
Robbery	6.8	11.7	15.6	6.3
Aggravated assault	11.3	19.9	20.6	14.5
Rape	1.1	1.4	1.7	1.2
Total violent	19.6	33.8	38.4	22.5
Burglary	20.6	18.2	38.5	17.5
Larceny	44.6	38.1	54.4	24.3
Motor vehicle theft	15.5	17.8	18.4	18.1
Total property	80.7	74.1	111.2	59.9
Total Index	100.4	107.9	149.6	82.4
	Rate per Square Mile			
	Citywide	North Target	East Target	Comparison Area
Murder	1.2	5.4	4.1	1.9
Robbery	27.6	69.5	135.5	25.7
Aggravated assault	45.4	118.3	178.1	59.3
Rape	4.5	8.2	14.8	4.8
Total violent	78.6	201.4	332.5	91.8
Burglary	82.6	108.6	333.7	71.1
Larceny	178.5	226.9	471.0	98.9
Motor vehicle theft	62.1	105.7	159.2	73.8
Total property	323.2	441.2	963.9	243.9
Total Index	401.9	642.6	1,296.4	335.6

Note: Numbers may not add up due to rounding.

The North District followed a more targeted deterrence approach. This selective approach to vehicle and pedestrian stops sought to maximize seizures of illegal weapons and drugs through more thorough, focused investigation. Resources were targeted toward individuals suspected of being involved in illegal activities.

Activity Data

In their effort to increase the visibility of the police and thus deter crime, the East target area used nearly 1,000 more officer hours than the North target area (see exhibit 3). East District police officers made more arrests, issued more warning tickets and traffic citations,[6] and seized more drugs.

But the North District's more selective, focused approach resulted in the detection of more serious criminal activity during each stop. As a result, North District officers made twice as many total arrests per vehicle stop. They were three times as likely to uncover an illegal firearm during a traffic stop and discovered three times as many total guns per stop.[7] (See exhibit 4.) As part of the strategy to target offenders and investigate suspicious behavior, North District officers also made 126 probation checks during proactive checks of probationers at their residences.

Exhibit 3 Directed Patrol Activity Data

Activity	Total Activities		
	Total	North Target Area	East Target Area
Officer hours	4,879.8	1,975	2,904.8
Traffic citations	1,638	698	940
Warning tickets	2,837	510	2,327
Combined tickets	4,475	1,208	3,267
Vehicle stops	5,253	1,417	3,836
Total arrests	992	434	558
Drug seizures	61	18	43
Illegal gun seizures	25	12	13
Legal guns discovered	81	43	38
Total guns	106	55	51

Exhibit 4 Comparison of Activity for Vehicle Stops

Activity	North Target per 100 Stops	East Target per 100 Stops
Total vehicle stops	1,417	3,836
Felony arrests	2.9	1.1
Total arrests	30.6	14.5
Illegal gun seizures	0.9	0.3
Total guns	3.9	1.3
Warning tickets	36.0	60.7
Citations	49.2	24.5
Probation contacts*	8.9	0.0

* Probation checks were based on addresses of probationers residing in the target beats rather than on routine vehicle stops.

Illegal Firearms Seized

Although the North District officers found more guns per vehicle stop, the total number of illegal firearms seized in the two target areas was similar. As exhibit 5 shows, 42 firearms were seized in the North target area and 45 were seized in the East target area during the 90-day directed patrol period in 1997. This compared with 39 and 30, respectively, during the same 90-day period in 1996. This represented a modest increase (7.7 percent) over 1996 levels in the North District and a sizable increase (50 percent) in the East District. In the comparison area, the number of seizures declined 40 percent from 1996 to 1997.

Observational Data

Trained observers rode with participating officers for 100 hours and observed 104 contacts between officers and citizens. According to the observers, East District officers were more likely to base their contact on a traffic law violation. In the North District, officers concentrated their efforts and contacts on suspicious persons or situations. Yet, when North District officers did make a traffic stop, they were more likely to issue a citation. In the North District, 26 percent of contacts resulted in traffic citations; in the East District, 9 percent of contacts resulted in citations. Contacts in both target areas lasted about 15 minutes.

Exhibit 5 Firearms Seized in Target and Comparison Areas During 90-Day Period in 1996 and 1997

Guns Seized*	North Target	East Target	Comparison Area
1996 (before directed patrol)	39	30	45
1997 (during directed patrol)	42	45	27
% change	+7.7	+50.0	−40.0

Note: Data are for the period from July 15 through October 15.
* Includes guns seized by directed patrol officers and regular duty officers.

Differences in Firearm-Related Crimes [8]

Although homicide declined in both North and East target areas (and the declines were significant when contrasted with the citywide trend), other gun-related offenses declined in the North target area but increased in the East target area (see exhibit 6). For example, aggravated assaults with a gun and armed robberies declined 40 percent in the North target area. These were statistically significant decreases compared with both the comparison area and the citywide trend. Similarly, total gun crimes declined 29 percent in the North target area. In contrast, aggravated assaults with a gun increased 58 percent and armed robberies increased 16 percent in the East target area. Although the increases in the East District were smaller than the increases observed in the comparison area, they contrast significantly with the decreases in the North target area.

Thus, other than for homicide, it appears that the positive effects on firearm-related crimes were confined to the North target area, where officers were more selective in who they stopped. Furthermore, decreases in robbery and aggravated assault rates in the North District were primarily due to the decline in firearm-related assaults and robberies.

Exhibit 6 Change in Firearm-Related Crime for 90-Day Period in 1996 and 1997

	North Target	East Target	Total Target Area	Comparison Area	Citywide
Homicide					
1996 (before directed patrol)	7	4	11	3	17
1997 (during directed patrol)	1	0	1	3	26
% change	NA*†	NA*‡	NA*†	NA*	+52.9
Aggravated assault—gun					
1996 (before directed patrol)	40	19	59	22	333
1997 (during directed patrol)	24	30	54	48	402
% change	−40.0≠	+57.9	−8.5ʃ	+118.2	+20.7
Armed robbery					
1996 (before directed patrol)	31	31	62	13	356
1997 (during directed patrol)	19	36	55	21	338
% change	−38.7#	+16.1	−11.3	+61.5	−5.0
Gun crimes					
1996 (before directed patrol)	75	42	117	49	NA
1997 (during directed patrol)	53	57	110	53	NA
% change	−29.3	+35.7	−6.0	+8.2	NA

Note: Data are for the period from July 15 to October 15. The homicide and armed robbery categories include some incidents that involved a weapon that was not a firearm.

* Percent change not calculated due to small total study population.

† Comparison to citywide trend significant at ≤ .05.

‡ Comparison to citywide trend significant at ≤ .10.

ʃ Comparison to citywide trend significant at ≤ .05; to comparison beats significant at ≤ .10.

≠ Comparison to both citywide trend and comparison beats significant at ≤ .05.

Comparison to comparison beats significant at ≤ .10.

Will the Community Support Aggressive Patrol Strategies?

Increasing the number of police officers on patrol, especially in a city's most disenfranchised neighborhoods, may increase citizens' feelings of safety and communication between the police and the community. But police managers also need to consider the possible adverse consequences of implementing aggressive patrol strategies. If citizens criticize the police and view the frequent stops as harassment, then any reduction in crime will be accomplished only with significant costs. Citizen support for the police may decrease, public criticism may increase, and racial tensions may intensify. These consequences, if they occur, would adversely affect any department's community policing program.

To learn how citizens perceive aggressive patrol strategies and investigate how an intense police presence affects citizens' opinions, researchers examined whether citizen perceptions of crime, fear, and disorder changed after the directed patrol program was implemented. Researchers also explored the community's awareness and acceptance of the program. Did the community support the aggressive effort, or did directed patrol increase the community's concerns about racial profiling and disparity in how the police treat people? Surveys were administered in both the target and comparison areas before and after directed patrol was conducted.[9] The findings: The community was aware of the program and supported it.

Awareness of and Support for Directed Patrol

Approximately two-thirds of citizens were aware of IPD's patrol strategies to remove guns and drugs from the streets (see note 9). Little change was found in the levels of awareness in the test and comparison samples. This suggests a general awareness of the strategies rather than an increased awareness as a result of the directed patrol experiment.

Before the directed patrol intervention, approximately 71 percent of the citizens in the sample supported "intense patrol" and increased police visibility. This figure increased to 76 percent following the program. Residents of the

East target area accounted for most of the increase. No change in support was found among North target or comparison area residents.

In the North District, the level of support from blacks was almost unchanged (from a mean score of 4.58 before the program to 4.62 after the program).[10] Whites in the North District were slightly less supportive, but their support increased more from the preprogram to the postprogram survey (4.09 to 4.25). In the East District, whites were slightly more supportive than blacks.

Whites in the North target area prior to directed patrol were the least supportive. Women were more supportive of directed patrol than were men, although the differences were not pronounced. No statistically significant changes in opinion by race or gender followed the directed patrol experiment. Thus, the effort neither built nor harmed public support.

Attitudes Toward the Police

Citizens also were asked about their general support for the police department. The high level of support for IPD prior to directed patrol did not change after the experiment ended. Whites expressed slightly more positive attitudes toward the police, although no statistically significant changes were found by race or sex in the postintervention survey.[11]

Minimizing Police-Citizen Conflict

Despite the large number of contacts between police and citizens, citations issued, and arrests made, IPD officials said that no reported citizen complaints were tied to the directed patrol initiative. IPD took several steps to prevent conflict from this aggressive police strategy. First, the deputy chief of each district attended community meetings and spoke with neighborhood leaders before directed patrol was implemented. The deputy chiefs explained the project and its goals and stated that the department would not implement the project if the community objected. Assured of neighborhood leaders' support, the deputy chiefs asked them to explain the project to residents and to solicit community support.

Second, the department provided adequate supervision for the project. A captain in each district was assigned to the project. A team of sergeants directly supervised the officers, often arriving on the scene of traffic stops and investigations. Furthermore, the captains and sergeants emphasized that the project had to be implemented in a way that was respectful of the citizens with whom officers had contact.

The citizen survey results suggest that IPD was successful in implementing the project in a fashion that did not generate police-citizen conflict.[12] The survey approach is unlikely, however, to tap into the perceptions of the most disenfranchised members of the community because they are unlikely to participate in a telephone survey. No evidence of such criticism exists, but it remains a possible effect of directed patrol efforts.

Impact on Perceptions of Crime, Quality of Life, and Fear

Although the changes were not dramatic, declines were seen in both the North and East target areas in the percentage of respondents indicating that neighborhood crime had increased. At the same time, a slight, although not significant, increase was seen in the comparison area. Drug and gun crimes were the highest rated crime problems in the target neighborhoods. Statistically significant declines were noted in the percentage of citizens in both target areas who labeled drugs as a "major problem" following directed patrol. Modest, although not statistically significant, declines also were observed for gun crime. No changes were observed in the comparison area.

North target area residents were less likely to rate their neighborhood negatively following directed patrol. Both North and East target area residents were less likely to claim that the neighborhood was a "worse place." Beyond this, however, no significant changes in perceptions of the neighborhood or fear of crime were noted.

Implications for Further Research

The results of this study indicate that directed patrol, using a **targeted** rather than a **broad** general deterrence strategy, can have a significant effect on violent crime. This finding is supported by the overall effect on homicide, the effect on firearm-related crime in the North target area (which used a targeted deterrence strategy), and the consistency with earlier findings in the Kansas City project. The less positive results on crime in the East target area, which used a broader general deterrence strategy, illustrate the greater effectiveness of a more focused deterrence approach.

Consequently, more needs to be learned about the effects of directed patrol strategies on crime. The Kansas City results show that removing illegal weapons from a high-crime neighborhood may be a key strategy in reducing firearm-related crime. The contrast between the North and East Districts, however, suggests that merely removing illegal firearms may not be the primary causal agent. Rather, a targeted deterrence approach that increases surveillance of suspicious individuals in high-risk neighborhoods may be the key ingredient to reducing gun crime and violence. More research is needed on the crime reduction effects of targeted versus more general deterrence approaches. Multiple-site, multiple-method tests of directed patrol interventions could be helpful in understanding the impact on different types of crime.[13]

Further Research on Directed Patrol and Police-Minority Community Relations

More needs to be learned about how to implement directed patrol projects while maintaining positive relationships with the community. Consistent findings emerge from Kansas City and Indianapolis about the impact these projects had on citizen perceptions of the police. Both the Kansas City target area and the North target area in Indianapolis were in predominantly black neighborhoods, involved aggressive patrol strategies, and received support from residents. The effort also was supported in the predominantly white neighborhoods in the East target area. Given the history of police-citizen relationships in the black community, it is striking to find high levels of support by blacks for an aggressive police strategy that can lead to significantly higher levels of vehicle stops by the police.

In their 1988 article, Sampson and Cohen quoted Lawrence W. Sherman:[14]

> Done properly, proactive strategies need not abuse minority rights or constitutional due process nor hinder community relations. But the difficulties of implementing such strategies are substantial, and great care is required at implementation. (Sherman, 1986: 379)

IPD district chiefs took the time to meet with neighborhood leaders and community groups to explain the initiative and secure their support before implementation. In addition, directed patrol supervisors emphasized the need to treat citizens with respect and explain to citizens why they were being stopped. Observations by trained observers suggested that officers' actions were consistent with these instructions. Beyond these points, however, more needs to be learned about the training and tactics that can be used to ensure that this type of aggressive strategy is positively received by the community. This point is given weight in the recent research reported by Paternoster and colleagues.[15] Although they looked specifically at arrests in spouse assault cases, researchers found that suspects' perception of the fairness of police treatment had long-term impacts on subsequent violence.

Turning Research Into Practice

Indianapolis's experience with directed patrol establishes the value of using research methods and data analysis to help solve crime problems. The study demonstrates the need to distinguish between two similar but distinct strategies that had very different levels of effectiveness in addressing gun-related violence.

The directed patrol experience helped Indianapolis officials understand the importance of linking research and practice. After the directed patrol experiment, it became more commonplace for officials to use research to enhance the success of strategic initiatives. One such initiative is a broad, multiagency problem-solving effort aimed at addressing homicide and gun violence. The initiative, known as the Indianapolis Violence Reduction Partnership (IVRP), includes local, State, and Federal criminal justice agencies; community leaders; service providers; and a research team. Researchers help IVRP pinpoint problems, develop a strategy, monitor how the strategy is working, interpret new data, and continually adjust the strategy as needed.

Analyses to date indicate that the IVRP approach resulted in significant reductions in homicide in Indianapolis during 1999.[16]

The link between research and practice also has produced promising results elsewhere. Boston's Operation Ceasefire, a communitywide collaboration, reduced youth homicide by more than 60 percent.[17] Other cities, including Indianapolis through IVRP, are adopting Boston's problem-solving model in the Strategic Approaches to Community Safety Initiative sponsored by the National Institute of Justice. The Boston and Indianapolis experiences, coupled with other promising interventions, suggest that serious crime problems can be productively addressed through these partnerships between criminal justice practitioners and researchers.[18]

Notes

1. Cordner, G.W., "The Effects of Directed Patrol: A National Quasi-Experiment in Pontiac," in *Contemporary Issues in Law Enforcement,* ed. J. Fyfe, Beverly Hills, CA: Sage Publications, 1981.

2. Sherman, L.W., D. Gottfredson, D. MacKenzie, J. Eck, P. Reuter, and S. Bushway, *Preventing Crime: What Works, What Doesn't, What's Promising,* Washington, DC: U.S. Department of Justice, National Institute of Justice, 1997, NCJ 165366.

3. Support for the deterrent effect of increasing the number of police and the contacts between police and citizens in high-crime areas is provided by Marvel, T.B., and C.E. Moody, "Specification Problems, Police Levels and Crime Rates," *Criminology* 34 (1996): 609–646; Wilson, J.Q., and B. Boland, "The Effect of the Police on Crime," *Law and Society Review* 12 (1978): 367–390; Sampson, R.J., and J. Cohen, "Deterrent Effects of the Police on Crime: A Replication and Theoretical Extension," *Law and Society Review* 22 (1988): 163–189; Press, S.J., *Some Effects of an Increase in Police Manpower in the 20th Precinct of New York City,* New York: New York City Rand Institute, 1971; and Schnelle, J.F., R.E. Kirchner, Jr., J.D. Casey, P.H. Uselton, Jr., and M.P. McNees, "Patrol Evaluation Research: A Multiple-Baseline Analysis of Saturation Police Patrolling During Day and Night Hours," *Journal of Applied Behavior Analysis* 10 (1977): 33–40.

4. The full technical report, including the time-series analysis discussed in the appendix, "Additional Methods and Findings," was submitted to the National Institute of Justice in 2000: McGarrell, E.F., S. Chermak, and A. Weiss, "Reducing Firearms Violence Through Directed Police Patrol: Final Report on the Evaluation of the Indianapolis Police Department's Directed Patrol Project." See also McGarrell, E.F., S. Chermak, A. Weiss, and J. Wilson, "Reducing Firearms Violence Through Directed Police Patrol," *Criminology & Public Policy* 1 (1) (November 2001): 119–148.

5. The targeted deterrence strategy was also evident in the Kansas City gun experiment. (Sherman, L.W., J.W. Shaw, and D.P. Rogan, *The Kansas City Gun Experiment,* Research in Brief, Washington, DC: U.S. Department of Justice, National Institute of Justice, 1995, NCJ 150855; and Sherman, L.W., and D.P. Rogan, "The Effects of Gun Seizures on Gun Violence: 'Hot Spots' Patrol in Kansas City," *Justice Quarterly* 12 (1995): 673–693.)

Similar targeted deterrence strategies were successful in San Diego and Boston. The San Diego field interrogation was intended to send a deterrence message to high-risk individuals in high-risk locations (Boydstun, J., *San Diego Field Interrogation: Final Report,* Washington, DC: Police Foundation, 1975). In Boston's Operation Ceasefire meetings, the threat of punishment was directly communicated to individuals believed to be at greatest risk for involvement in firearm-related violence (Kennedy, D.M., "Pulling Levers: Getting Deterrence Right," *National Institute of Justice Journal* 236 (July 1998): 2–8). See also Sherman, L.W., "Policing for Crime Prevention," in *Preventing Crime,* by Sherman, Gottfredson, MacKenzie, Eck, Reuter, and Bushway (see note 2).

6. North and East District officers issued 1,638 traffic citations and 2,837 warning tickets. Officers said that warning tickets often were issued for minor infractions because issuing expensive tickets to low-income residents frustrates their efforts to cultivate positive community relations. Citations were issued for more serious infractions and repeat violations. The directed patrol experiment in Indianapolis resulted in 84 felony arrests, 654 misdemeanor arrests, and 254 warrant arrests, for a total of 992 arrests.

7. Directed patrol officers seized 25 illegal firearms; an additional 81 legally possessed weapons were discovered. (Additional firearms were seized in the target areas by officers on routine patrol. Exhibits 4 and 5 reflect the total firearms seized.) Thus, officers uncovered more than three legally possessed weapons for every one illegally possessed weapon. Some officers joked that people who were stopped were more likely to have a gun permit than a driver's license.

8. Exhibit 6 reports the results of statistical significance tests conducted using the general linear model analysis of variance approach. The variance is partitioned into period effects, area effects, and effects due to the interaction of area and period. The interaction effect is of interest because it allows the study to contrast the trend in the target areas with the trend in the comparison area and in the city as a whole. When the target areas experience a decline in crime, the method tests whether the decline is greater than would be expected by chance given the trend in the comparison area.

Similarly, when the target areas experience no change or an increase, the method allows the study to test whether it is significantly different from the trend in the comparison area.

9. The surveys were conducted using a randomly selected panel design. The sample size of the preintervention panel was 420. The net response rate was 58.2 percent (302 of the 722 people approached refused to participate). The response rates in the North (59.7 percent) and East (66.5 percent) target areas were higher than in the comparison area (49.6 percent). The sample size of the panel following the directed patrol experiment was 282 (67 percent of the original sample). Attrition was evenly distributed throughout the two target areas and the comparison area and the demographic patterns were consistent in both the presamples and postsamples. Additional information on the survey is available in Chermak, S., and E.F. McGarrell, "Citizens' Perceptions of Aggressive Traffic Enforcement Strategies," *Justice Quarterly* 18 (2001): 365–391.

10. Support for the program was measured on a five-point Likert scale from "no support" to "strongly support."

11. A small decline occurred in the number of North District residents who strongly agreed that police were professional and courteous. This was not found in the East District, but it was observed in the comparison area. Thus, it does not seem to be the product of directed patrol. The decline most likely is the result of the highly publicized trial of four IPD officers who were involved in a confrontation with two citizens while the officers were off duty. The trial occurred during the directed patrol experiment. This appeared to affect attitudes of black citizens, who are the majority of citizens in the North District and comparison area.

12. See also Shaw, J.W., "Community Policing Against Guns: Public Opinion of the Kansas City Gun Experiment," *Justice Quarterly* 12 (1995): 695–710.

13. Sherman, Gottfredson, MacKenzie, Eck, Reuter, and Bushway, *Preventing Crime* (see note 2); and Sherman, L.W., "Cooling the Hot Spots of Homicide: A Plan for Action," in *What Can the Federal Government Do to Decrease Crime and Revitalize Communities?* Research Forum, Washington, DC: U.S. Department of Justice, National Institute of Justice and Executive Office for Weed and Seed, 1998, NCJ 172210.

14. Sampson and Cohen, "Deterrent Effects of the Police on Crime" (see note 3).

15. Paternoster, R., R. Bachman, R. Brame, and L.W. Sherman, "Do Fair Procedures Matter? The Effect of Procedural Justice on Spouse Assault," *Law and Society Review* 31 (1997): 163–204.

16. Coleman, V., W.C. Holton, Jr., K. Olson, S.C. Robinson, and J. Stewart, "Using Knowledge and Teamwork To Reduce Crime," *National Institute of Justice Journal* 241 (October 1999): 16–23; and McGarrell, E.F., and S. Chermak, "Problem Solving to Reduce Gang and Drug-Related Violence in Indianapolis," in *Gangs, Youth Violence, and Community Policing,* edited by S. Decker, Belmont, CA: Wadsworth, forthcoming.

17. See Kennedy, D.M., A.A. Braga, A.M. Piehl, and E.J. Waring, *The Boston Gun Project's Operation Ceasefire,* Reducing Gun Violence Series, Washington, DC: U.S. Department of Justice, National Institute of Justice, 2001, NCJ 188741.

18. The data-driven problem-solving model is a key component of the U.S. Department of Justice's Project Safe Neighborhoods (visit the Project Safe Neighborhoods Web site at *http://www.psn.gov)*.

Appendix: Additional Methods and Findings

To further test the potential impact on violent crime, three ARIMA (auto-regressive integrated moving average) models were used to compare the North and East target areas, the comparison area, and the city (minus the target areas). The outcome measure was the number of homicides, aggravated assaults with a gun, and armed robberies. The data were compiled from the first week in 1995 through January 12, 1998—a 132-week preintervention period, a 13-week intervention period, and a 13-week postintervention period.

The first model compared the three periods. A significant effect was found for the North target area, which had approximately two fewer violent crimes on average during the intervention. The comparison area witnessed an increase of slightly more than one violent crime per week on average during the intervention period. Neither the East target area nor the net city-wide trend showed significant changes during the intervention period.

The second model compared the intervention period with the preintervention period. An effect of reduced violent crimes was found in the North target area during the intervention period. The comparison area witnessed an increase, whereas neither the East target area nor the city trend saw a change.

The third model compared the intervention period with the postintervention period. This tested the effect of removing the intervention. No significant changes were found when the intervention was removed. This suggests that the intervention effect in the North target area remained even after the directed patrol project ended. This model should be interpreted cautiously, however, because of the small number of observations (26 weeks).

It appears unlikely that the findings from the East target area were due either to a long-term suppression effect or a rebound from unusually low levels of violent crime.

To determine whether evidence of a residual deterrence effect existed, researchers compared the trend in crime for the 90-day period following the termination of directed patrol (October 16, 1997, to January 15, 1998) with the same period in the previous year. The findings from this analysis are mixed. Although homicides continued to decline in the North target area, they increased in the East target area. Aggravated assault with a gun declined 30 and 49 percent in the North and East target areas, respectively; declines also occurred in the comparison area and citywide, although they were of a smaller magnitude than in the target areas. The differences between the target areas and the comparison area were not statistically significant. Armed robberies declined 15 percent in the North target area, similar to the citywide trend. Both the East target area and the comparison area witnessed increases in armed robbery. Thus, the most promising results are for aggravated assault with a gun. Both target areas experienced fairly large decreases, although the lack of statistical significance when contrasted with the comparison area suggests that these results reflect the citywide trend rather than the residual deterrent effect of directed patrol. The findings for other offenses did not reveal any consistent evidence of residual deterrence.

About the National Institute of Justice

NIJ is the research, development, and evaluation agency of the U.S. Department of Justice and is solely dedicated to researching crime control and justice issues. NIJ provides objective, independent, nonpartisan, evidence-based knowledge and tools to meet the challenges of crime and justice, particularly at the State and local levels. NIJ's principal authorities are derived from the Omnibus Crime Control and Safe Streets Act of 1968, as amended (42 U.S.C. §§ 3721–3723).

The NIJ Director is appointed by the President and confirmed by the Senate. The NIJ Director establishes the Institute's objectives, guided by the priorities of the Office of Justice Programs, the U.S. Department of Justice, and the needs of the field. The Institute actively solicits the views of criminal justice and other professionals and researchers to inform its search for the knowledge and tools to guide policy and practice.

NIJ's Mission

NIJ's mission is to advance scientific research, development, and evaluation to enhance the administration of justice and public safety.

NIJ's Strategic Goals and Program Areas

NIJ has seven strategic goals grouped into three categories:

Creating relevant knowledge and tools:

1. Partner with State and local practitioners and policymakers to identify social science research and technology needs.
2. Create scientific, relevant, and reliable knowledge—with a particular emphasis on violent crime, drugs and crime, cost-effectiveness, and community-based efforts—to enhance the administration of justice and public safety.
3. Develop affordable and effective tools and technologies to enhance the administration of justice and public safety.

Dissemination:

4. Disseminate relevant knowledge and information to practitioners and policymakers in an understandable, timely, and concise manner.
5. Act as an honest broker to identify the information, tools, and technologies that respond to the needs of stakeholders.

Agency management:

6. Practice fairness and openness in the research and development process.
7. Ensure professionalism, excellence, accountability, cost-effectiveness, and integrity in the management and conduct of NIJ activities and programs.

NIJ's Structure

NIJ has three operating units. The Office of Research and Evaluation manages social science research and evaluation and crime mapping research. The Office of Science and Technology manages technology research and development, standards development, and technology assistance to State and local law enforcement and corrections agencies. The Office of Development and Communications manages field tests of model programs, international research, and knowledge dissemination programs. NIJ is a component of the Office of Justice Programs, which also includes the Bureau of Justice Assistance, the Bureau of Justice Statistics, the Office of Juvenile Justice and Delinquency Prevention, and the Office for Victims of Crime.

To find out more about the National Institute of Justice, please contact:

National Criminal Justice Reference Service
P.O. Box 6000
Rockville, MD 20849–6000
800–851–3420
e-mail: *askncjrs@ncjrs.org*

To obtain an electronic version of this document, access the NIJ Web site
(*http://www.ojp.usdoj.gov/nij*).

If you have questions, call or e-mail NCJRS.

U.S. Department of Justice

Office of Justice Programs

National Institute of Justice

Washington, DC 20531

Official Business

Penalty for Private Use $300

N C J   1 8 8 7 4 0

PRESORTED STANDARD
POSTAGE & FEES PAID
DOJ/NIJ
PERMIT NO. G–91

www.ingramcontent.com/pod-product-compliance
Lightning Source LLC
Chambersburg PA
CBHW081242170526
45165CB00009B/3162